ASSESSMENT ALTERNATIVES
in MATHEMATICS

An overview of assessment techniques that promote learning

Prepared by the EQUALS staff
and the Assessment Committee of the
California Mathematics Council
Campaign for Mathematics

Jean Kerr Stenmark

Illustrations:
Adapted from:
EQUALS publications
Math for Girls and Other Problem Solvers,
FAMILY MATH, Get It Together, SPACES,
Off and Running,
and unpublished student writing;
from Shell Centre Publications
The Language of Functions and Graphs,
and *Problems with Patterns and Numbers;*
from California Assessment Program booklets
A Question of Thinking,
Survey of Basic Skills,
and teacher comments on portfolios;
and from National Assessment of Educational
Progress publication
Learning by Doing.

For information about additional copies, contact:
EQUALS
Lawrence Hall of Science
University of California
Berkeley, California 94720
Attention: *Assessment Booklet*

**Co-sponsored by EQUALS and the California Mathematics
Council** *Campaign for Mathemtics*
The *Campaign for Mathematics* is funded by a grant from the
California Postsecondary Education Commission and the
California State Department of Education under Title II of Public
Law 98-377, now called the Eisenhower program.
The publications of EQUALS are partially supported by a grant
from the Carnegie Corporation of New York.

10 9 8 7

ISBN 0-912511-54-0

ASSESSMENT ALTERNATIVES IN MATHEMATICS
TABLE OF CONTENTS

PREFACE

This publication combines resources of the California Mathematics Council and the EQUALS program of the Lawrence Hall of Science at the University of California, Berkeley.

The EQUALS program has for many years been concerned with assessment as an equity issue, and has urged educators to explore means of assessing other than through multiple-choice tests.

In October, 1988, The CMC *Campaign for Mathematics*, funded by a grant from the California Postsecondary Education Commission, held its first convocation in Beaumont, California. The Assessment Committee discussed many assessment ideas and ways to communicate with educators and the public about assessment.

A plan was made for CMC and EQUALS to collaborate in writing a booklet about alternatives to standardized testing. It was agreed that EQUALS staff would coordinate the writing, that CMC would pay most of the printing costs, and that "ownership" of the resulting booklet would be shared.

This is the resulting booklet. It is being distributed to all CMC members as part of the December, 1989, issue of *The CMC ComMuniCator* and will be distributed at other CMC functions. The EQUALS program will reprint the book and have it available for sale in the future.

Members of the Campaign Assessment Committee were Joan Akers, Sandy Bright, Elaine Harvey, Elisabeth Javor, Peggy Kelly, Martin Lang, Alfred Manaster, Judy Mumme (Chair), Susan Ostergard, Tej Pandey, Barbara Pence, Jean Stenmark, and Dorothy Wood. Phil Daro, Walter Denham, and Jane Stenmark have also helped with comments and feedback.

I would also like to thank other members of the EQUALS staff, Nancy Kreinberg, Lynne Alper, Sue Arnold, Mary Jo Cittadino, Ruth Cossey, Tim Erickson, Sherry Fraser, Kay Gilliland, Thalia Guy, Ellen Humm, Helen Joseph, Helen Raymond, Celia Stevenson, Virginia Thompson, and Linda Witnov, for their helpful editing and other support.

Please note that this booklet is a beginning, not an end. It barely touches the surface of the possibilities to make assessment support student learning. More could and should be said about any issue, and more issues will arise. We hope these pages will spark many discussions about the revolution going on in the world of testing and assessment, in classrooms and at district, state, national, and international levels.

Jean Kerr Stenmark

September, 1989

Wordsmith

MATHEMATICS ASSESSMENT ALTERNATIVES

Introduction	THIS BOOKLET is about mathematics assessment alternatives and how they can be used in California. It is meant for teachers, administrators, parents, and anyone else who wants a brief overview of current trends in assessment.
The Vision for Mathematics Education	Some mathematics educators have this vision of what should be happening in a mathematics class: Students work in small groups or independently, doing investigations and projects, using tools such as manipulative materials, calculators, computers, assorted textbooks, and other reference books. They consult with each other and with the teacher, keeping journals and other written reports of their work. Occasionally the entire class is called together for a discussion or meeting. The curriculum is rich in real problem solving and includes a full range of mathematical ideas. Mathematical power is at work in every part of this classroom. Do our present assessment methods support this vision?
Assessment in Classrooms	The first use of assessment, of course, is within the classroom, to provide information for making instructional decisions. Teachers have always depended on their own observations and examination of student work to assist in curriculum design and decision-making.
External Assessment Needs to Be Consistent with Curriculum Goals	But it is equally important that those outside the classroom have information more directly related to the vision than the abstract and meaningless numbers provided by standardized tests. Both internal and external assessment should be consistent with new curriculum standards. Classroom teachers need support in their efforts to set high goals for student achievement.
Real World Assessment	In the world of work, people are valued for the tasks or projects they do, their ability to work with others, and their responses to problem situations. To prepare students for future success, both curriculum and assessment must promote this kind of performance.
New Assessment Techniques	This document provides an overview of some possible assessment methods, most of which are already used effectively in various places. Although some of these methods represent significant changes from present common practices in this country, we see that many groups of educators around the world are exploring and creating exciting new possiblities.
Terminology	Readers will not need to be especially well-informed about current testing terminology, such as the difference between standardized and normed tests, evaluation versus assessment, or the importance of p-values and standard deviation. For the language of present testing programs, you may want to consult an evaluation and statistics textbook.

This is a page from *Assessment Alternatives in Mathematics*, a booklet from the California Mathematics Council and EQUALS.

MATHEMATICS ASSESSMENT ALTERNATIVES (continued)

Authentic Achievement

We present in this booklet some ways to bring about assessment of authentic achievement. The implications are that students should be working on worthwhile investigations or tasks and that their success should be evaluated in ways that make sense.

Issues

Not every aspect of assessment is included and much of what we have included is incomplete. There are many issues to be addressed. Some are briefly discussed on pages 31-34, and others will no doubt arise in your mind as you read.

Arrangement of Sections

There is much overlap among the sections of this book; you will see the same ideas repeated several times. Portfolios, for instance, might include something from each of the other assessment methods. A performance task might be part of an investigation.

It is important to remember that what once might have been considered "classroom assessment" will now be included often in "external assessment" as we move toward a better match between the two reasons for assessing.

There are four sections of the booklet:
> **Introductory** section, on pages 1 through 5
> **Assessment of products** that students generate, pages 6 through 19, including portfolios, writing, investigations, and open-ended questions
> **Assessment of the process** students are using, pages 20 through 25, including observations, interviews, and questions
> **Other matters**, pages 26 through 35, including student self-assessment, sample problems, the California Assessment Program, issues in assessment, and suggestions for action.

Additional Information

Sources of quotes and of additional information are given, including addresses where available, on pages 36-38. Some of the listed publications are not referred to in the text but are good resources.

We hope you will find the information of the booklet valuable. The decade between 1990 and 2000 should see momentous changes in assessment in California.

As we need standards for curricula, so we need standards for assessment. We must ensure that tests measure what is of value, not just what is easy to test. If we want students to investigate, explore, and discover, assessment must not measure just mimicry mathematics. By confusing means and ends, by making testing more important than learning, present practice holds today's students hostage to yesterday's mistakes. (Everybody Counts, p. 70)

This is a page from *Assessment Alternatives in Mathematics*, a booklet from the California Mathematics Council and EQUALS.

ASSESSMENT SHOULD PROMOTE LEARNING

Purpose of Assessment

The purpose of assessment should be to improve learning. With this in mind, we focus in this booklet on expanding how students can demonstrate their mathematical achievements and how teachers can gain better information about their students. This way of looking at learning means that there will be less need for complex scoring or grading of student work. Comparing students will become less important than helping students understand mathematics.

What Kind of Assessment Do We Need in Mathematics?

We need mathematics assessment that:

- matches the ideal curriculum described in such documents as the *California Mathematics Framework* and the *NCTM Standards* in both what is taught and how it is experienced, with thoughtful questions that allow for thoughtful responses
- communicates to students, teachers, and parents that most real problems cannot be solved quickly and that many have more than one answer
- allows students to learn at their own pace
- focuses on what students do know and can do rather than what they don't know
- won't label half of students as failures because of unrealistic expectations that all scores should be above the 50th percentile
- doesn't use time as a factor, since speed is almost never relevant in mathematical effectiveness
- is integral to instruction and doesn't detract from students' opportunities to continue to learn

What Do We Want to Assess?

We want to assess such things as:

- students' use of mathematics to make sense of complex situations
- students' work on extended investigations
- the ability of students to:
  - formulate and refine hypotheses
  - collect and organize information
  - explain a concept orally or in writing
  - work with poorly defined problems or problems with more than one answer, similar to those in real life
- students' use of mathematical processes, such as computation, in the context of many kinds of problems rather than in isolation
- the extent of students' understanding or misunderstanding about mathematical concepts
- students' ability to define and formulate problems
- whether students question possible solutions, looking at all possibilities
- how a student's productive work changes over time

This is a page from *Assessment Alternatives in Mathematics*, a booklet from the California Mathematics Council and EQUALS.

ASSESSMENT SHOULD PROMOTE LEARNING (continued)

What Kind of Activity Should We See?

We should see students:

- using mathematics with facility to communicate their own thinking about complex situations through pictures, diagrams, graphs, words, symbols, or numerical examples
- solving problems using a variety of mathematical tools and models, such as manipulatives, calculators, and computers
- planning, inventing, designing, and evaluating their own mathematical ideas and products
- being thoughtful, persistent, flexible, self-directed, and confident
- doing projects and activities
- working well together developing group problem-solving skills
- taking pleasure in doing mathematics

It is essential that internal assessment for instructional decisions and external assessment for other purposes be in agreement and that assessment always promote student learning. Beginning with the discussion of student products on the following page, this booklet presents many ideas for authentic assessment.

In today's political climate, tests are inadequate and misleading as measures of achievement. Assessment tasks should be redesigned—indeed, are being redesigned—to more closely resemble real learning tasks. Tests should require more complex and challenging mental processes from students. They should acknowledge more than one approach or one right answer and should place more emphasis on uncoached explanations and real student products. (Shepard, Why We Need Better Assessments, Educational Leadership, pp. 6-7)

This is a page from *Assessment Alternatives in Mathematics*, a booklet from the California Mathematics Council and EQUALS.

STUDENT MATHEMATICAL PRODUCTS

Beyond school, we demonstrate knowledge by providing original conversation and writing, by repairing and building physical objects, and by producing artistic, musical, and athletic performances.
In contrast, assessment in school usually asks students to identify the discourse, things, and performances that others have produced...
(Archbald, Beyond Standardized Testing, p. 3)

What are Student Products?

Student products are work that students have generated. Products may include writing in the form of journals or open-ended questions, videotapes, audiotapes, computer demonstrations, dramatic performances, bulletin boards, debates, student conference presentations, student designs and inventions, investigation reports, simulations, mathematical art, physical constructions of mathematical models — the list is almost endless.

Why Student Products?

Some possible objectives of student products indicate the variety of advantages in their use for assessment. Students can demonstrate:

- understanding of the mathematical ideas involved
- originality that goes beyond what had been taught
- ability to present reports, either orally or in writing, in an effective and attractive manner
- growth in social and academic skills and attitudes that will not be reflected in standardized tests
- success in meeting criteria determined ahead of time by teacher and students

More Advantages

Besides those indicated above, some other advantages of using student products for instruction and assessment:

- engaging students who are not enthusiastic about school
- bringing education to life, making it memorable for students
- demonstrating to the community what students are achieving, in a very real way
- providing a bridge between classroom and real world activities
- allowing for integration of mathematics with other subject areas
- giving students more flexible time to do thoughtful work
- permitting students to work with others
- encouraging creativity

Assessment Methods

The means of assessing student products may be as varied as the products themselves. Some suggestions:

- collect examples of many kinds of student work in portfolios
- decide on standards before projects are started, but keep the standards flexible enough to allow for the unusual

This is a page from *Assessment Alternatives in Mathematics*, a booklet from the California Mathematics Council and EQUALS.

STODENT MATHEMATICAL PRODUCTS (continued)

- involve community members and other students in evaluating student projects
- give comments about the work rather than a number rating or letter grade
- look for unique aspects or features that can be recognized for each product — make the results positive for all students
- consider including photographs of student products or students in action in student portfolios as a means of recording
- if grades or other comparisons are needed, choose holistic "rating" (see page 17 for more discussion of holistic scoring) rather than assigning points for small parts of the projects

The following pages include examples of student products in the form of portfolios, practical and mathematical investigations, writing in mathematics, and open-ended questions. We stress again that there are many other possibilities for educators to consider. This is only a sample.

STUDENT PORTFOLIOS

Mathematics Portfolios

Student portfolios are well-known in art and writing, but until now have rarely been used to keep a record of student progress in mathematics. Teachers have always kept folders of student work, but portfolios may now have more focus and be more important for assessment.

What is in a Portfolio?

Teachers and their students should be allowed to choose most of the items to include in portfolios, since it gives a good indication of what is valued. Occasionally it may be desirable, for the sake of comparisons, for some outside agency to ask for inclusion of a certain type of item, but this should be the exception. If possible, teachers and students should be able to present and explain their own portfolios to outside observers.

Putting dates on all papers will become more important. First draft or revised writing should be acceptable, but with a note about which it is. The names of group members should probably be on papers done by a group, or at least an indication that it was group work.

A portfolio might include samples of student-produced:

- written descriptions of the results of practical or mathematical investigations
- pictures and dictated reports from younger students
- extended analyses of problem situations and investigations
- descriptions and diagrams of problem-solving processes
- statistical studies and graphic representations
- reports of investigations of major mathematical ideas such as the relationship between functions, coordinate graphs, arithmetic, algebra, and geometry
- responses to open-ended questions or homework problems
- group reports and photographs of student projects
- copies of awards or prizes
- video, audio, and computer-generated examples of student work
- other material based on project ideas developed with colleagues

Teachers and Portfolios

The definition and evaluation of portfolios are opportunities for teachers to share and learn with peers. Groups of teachers who have reviewed the contents together have found it an exciting and rewarding experience. On page 10 are some examples of teacher comments made during pilot assessments in the spring of 1989. Also, sharing with parents, administrators, and school boards will help emphasize student accomplishments.

This is a page from *Assessment Alternatives in Mathematics*, a booklet from the California Mathematics Council and EQUALS.

STUDENT PORTFOLIOS (continued)

Advantages of Portfolios

Student portfolios can provide:

- evidence of performance beyond factual knowledge gained
- assessment records that reflect the emphases of a good mathematics program
- a permanent and long-term record of a student's progress, reflecting the life-long nature of learning
- a clear and understandable picture, instead of a mysterious test score number
- opportunities for improved student self-image as a result of showing accomplishments rather than deficiencies
- recognition of different learning styles, making assessment less culture dependent and less biased
- an active role for students in assessing and selecting their work

Student Attitudes

A portfolio may also incorporate important information about student attitudes toward mathematics, such as:

- a mathematical biography, renewed each year
- student self-report of what has been learned and/or what is yet to be learned
- a description of how the student feels about mathematics
- work of the student's own choosing
- excerpts from a student's mathematics journal

Assessment of Portfolios

Educators should look at many portfolios before trying to establish a standard of assessment. Because portfolios should reflect the instructional goals of each situation, the "rubrics" (detailed descriptions of assessment standards) will vary.

PORTFOLIOS:
Teacher Comments from CAP Pilot Project, Spring, 1989

The portfolio process has encouraged me to continue implementing all of the strands of the framework in my classroom. It gave me a lot of insight into individual student growth, understanding of mathematical concepts and "what" the students were thinking about and learning. I felt that I could get into their heads when I read their written pieces. I found out what activities they enjoyed because third graders are extremely honest. This process forced me to continue teaching rather than just "preparing" for a test. I feel that my students have gained a tremendous amount mathematical confidence as a result of their participation in this process

The portfolio process was a very positive experience in my classroom. It helped me draw several ideas together that might have languished awhile otherwise. I felt it gave validity to much of the work we have been doing in class.

The children were excited about sharing projects they were proud of with someone else (audience). We talked a lot about communication results and conclusions with others. Next time I will include them more in choosing projects

The big winner in this process is me! I have gained many ideas for improving instructional delivery during this process. My students need to keep daily journals while working on cooperative projects. They work at many levels while at the same project — that would help me track individual growth. Photos of large projects make sense. I would like to keep portfolios for conferences — picking pieces to exhibit, etc. I see

WRITING IN MATHEMATICS

Writing Requires Understanding

Communication in mathematics has become important as we move into an era of a "thinking" curriculum. Students are urged to discuss ideas with each other, to ask questions, to diagram and graph problem situations for clarity. Writing in mathematics classes, once rare, will now be vital.

California Students Need to Write

A review of the California State Department of Education's report on open-ended questions, *A Question of Thinking*, shows that most students lack opportunities to express mathematical ideas in writing, with fewer than 25% able to write competently about any of the problems given.

Forms of Writing in Mathematics

Mathematics writing may take many forms. It may be a separate activity, or may be a part of a larger project. Journals, reports of investigations, explanations of the processes used in solving a problem, or responses to open-ended questions all become part of what students do daily in classrooms as well as what is reviewed for assessment purposes.

At all levels including primary, writing or dictating to an adult can help students see connections. For example, in one of the portfolios we have seen, young students had made a network of all the words they could think of regarding "multiplication."

LIKE ADDING

NUMBERS GET BIGGER

OPPOSITE OF DIVISION

MULTIPLICATION

HARD

HOMEWORK

Feedback

Nahrgang (*Using Writing to Learn Mathematics*, p. 461) points out that writing enhances learning by giving "students the opportunity to formulate, organize, internalize, and evaluate" concepts. However, the benefits may not be automatic. Feedback and self-assessment are vital. Stream-of-consciousness writing while a problem is being solved provides a good record of thinking, but to be able to make generalizations students need to review, re-organize, and re-write.

Two-Step Assessment

Some written assessment, for example, is of a two-step form. The student writes about a problem or investigation, hands it in or discusses it with other students, then revises. The revised version becomes the official paper for evaluation.

Investigations and Open-ended Questions

Writing is inevitably woven into most of the assessment methods discussed in this booklet. In the following two sections, mathematical investigations and open-ended questions are presented as examples of use of mathematical writing.

SAMPLE STUDENT WRITING

This week our problem was to find out how many bananas Cori the camel could take to market, which was 1,000 miles away. Cori was kind of big but could only carry 1,000 bananas at a time. Does this problem so far sound easy? Well, its not! Corie also eats a banana every mile. It does not matter if she runs or walks backwards she still eats a banana. Yes, Cori walks she doesn't fly or have any helpers. ⌇⌇⌇⌇
~A Clue!~

Cori brings more than 10 bananas to the market.

My Approach

My approach was to draw a picture kind of like this;

First I tried having Cori walk 500 miles then walk back and get another 1,000 bananas but that didn't work. So next I tried having her take 1,000 bananas and walking 400 miles then have her leave 200 there and walk back to get another 1,000 bananas. By the time she got back to where she left her other 200 bananas she had 800 bananas and only 600 more miles to walk. So she had 200 bananas to sell at the market. There are more ways I'm sure but I think this one will do.

I liked this problem more than any other problem we have done, probobly because it had to do with camels. I love camels! ☺

What I learned from this problem was that I never will buy an oasis in the desert, 1,000 miles away from a market without a plain. Also why didn't Cori just eat her bananas instead of selling them, because even if she made money from the bananas its 1,000 miles away from the market which would be where she bought things.

This problem analysis was written by a seventh-grade student.

INVESTIGATIONS IN MATHEMATICS

Combining Instruction with Assessment

One of the best ways to assure the connection between instruction and assessment is to imbed assessment into instruction. When students become involved in practical or mathematical investigations, assessment can become natural and invisible. During the investigation, assessment activities or questions can be presented to the students without their being aware of any difference between the assessment and other classroom work.

Integrated Curriculum

Investigations may be related to other subject areas, such as science or social studies, or they may be explorations of purely mathematical questions.

Combining Several Modes of Assessment

Although the most typical form of assessment is collection of student writings, diagrams, graphs, tables, or charts, there are also opportunities for observation or videotaping of student performance.

Investigation Reviews

In reviewing investigations, look for whether students can:
- identify and define a problem and what they already know
- make a plan, creating, modifying, and interpreting strategies
- collect needed information
- organize the information and look for patterns
- discuss, review, revise and explain results
- persist, looking for more information if needed
- produce a quality product or report

Examples from Great Britain

On the following pages are some examples, from the Shell Centre in Great Britain, of questions and responses to assessment questions from an instructional/assessment module, and one marking scheme.

The opportunities for instruction and assessment of this type are endless. Here is a short list of mathematics-related investigations that would allow for inclusion of assessment . See pages 24-25 for a list of typical questions that may help elicit information about student understanding of concepts.

A Few Possible Investigations:

- Gears and ratios
- Maps
- Sound waves and music
- Traffic patterns near the school
- What is a safe distance from the car in front?
- Collecting and analyzing litter
- Study of local water use
- Population and availability of resources
- Census studies

- Use of tools
- Sports statistics
- Classroom Olympics
- Location of sunrise, sunset, moonrise, moonset
- Plant growth
- Measurement of parts of the body
- Diet, exercise, and health
- Comparisons of area and perimeter, surface area and volume

This is a page from *Assessment Alternatives in Mathematics*, a booklet from the California Mathematics Council and EQUALS.

SAMPLE ASSESSMENT ITEMS from Shell Centre Modules

THE CASSETTE TAPE

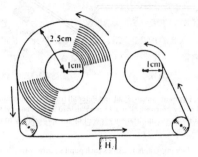

This diagram represents a cassette recorder just as it is beginning to play a tape. The tape passes the "head" (Labelled H) at a constant speed and the tape is wound from the left hand spool on to the right hand spool.

At the beginning, the radius of the tape on the left hand spool is 2.5 cm. The tape lasts 45 minutes.

(i) Sketch a graph to show how the *length* of the tape on the left hand spool changes with time.

Length of tape on left hand spool

Time (minutes)

THE CASSETTE TAPE (continued)

(ii) Sketch a graph to show how the *radius* of the tape on the left hand spool changes with time.

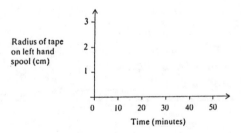

Radius of tape on left hand spool (cm)

Time (minutes)

(iii) Describe and explain how the radius of the tape on the *right-hand* spool changes with time.

_____
_____
_____
_____
_____
_____
_____
_____
_____

THE CASSETTE TAPE...MARKING SCHEME

(i) and (ii)   Translating words and pictures into mathematical representations.

(i)   *1 mark*   for a sketch graph showing a straight line with a negative gradient.

   1 mark   for a sketch ending at (45,0).

(ii)   *1 mark*   for a sketch beginning at (0,2.5) and ending at (45,1).

   1 mark   for a sketch showing a curve.

   1 mark   for a curve that is concave downwards.

(iii)   Describing and explaining a functional relationship using words.

   2 marks   for a correct, complete description.
   eg: 'the radius increases quickly at first, but then slows down'.

   Part mark: Give 1 mark for 'the radius increases'.

   2 marks   for a correct, complete explanation.
   eg: "the tape goes at a constant speed, but the circumference is increasing" or "the bigger the radius, the more tape is needed to wrap around it".

   Part mark: Give 1 mark for an explanation that is apparently correct but not very clear.

A total of 9 marks are available for this question.

Some Student Responses:

Stephanie

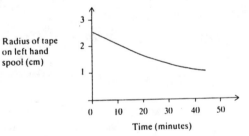

Radius of tape on left hand spool (cm)

Time (minutes)

Stephanie's sketch shows a curve beginning at (0,2.5) and ending at (45,1).

Brian

At the start it will have a radius of ∅. As the circumference is getting larger the amount of tape needed to go round it increases and as the tape is going past the head at a constant speed the radius will increase at a high speed at first but gradually slow down

This is a page from *Assessment Alternatives in Mathematics*, a booklet from the California Mathematics Council and EQUALS.

THE JOURNEY

The map and the graph below describe a car journey from Nottingham to Crawley using the M1 and M23 motorways.

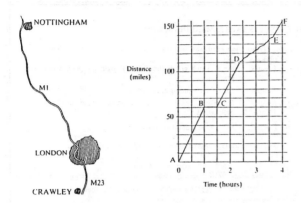

(i) Describe each stage of the journey, making use of the graph *and* the map. In particular describe and explain what is happening from A to B; B to C; C to D; D to E and E to F.

(ii) Using the information given above, sketch a graph to show how the speed of the car varies during the journey.

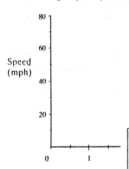

GOING TO SCHOOL

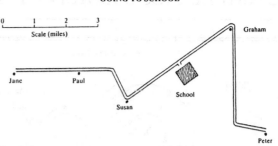

Jane, Graham, Susan, Paul and Peter all travel to school along the same country road every morning. Peter goes in his dad's car, Jane cycles and Susan walks. The other two children vary how they travel from day to day. The map above shows where each person lives.

The following graph describes each pupil's journey to school last Monday.

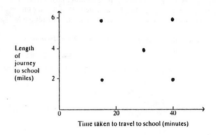

i) Label each point on the graph with the name of the person it represents.

ii) How did Paul and Graham travel to school on Monday? _____

_____

iii) Describe how you arrived at your answer to part (ii) _____

_____

_____

(continued)

CAMPING

On their arrival at a campsite, a group of campers are given a piece of string 50 metres long and four flag poles with which they have to mark out a rectangular boundary for their tent.

They decide to pitch their tent next to a river as shown below. This means that the string has to be used for only three sides of the boundary.

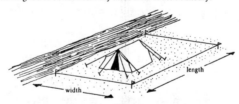

(i) If they decide to make the width of the boundary 20 metres, what will the length of the boundary be?

(ii) Describe in words, as fully as possible, how the length of the boundary changes as the width increases through all possible values. (Consider both small and large values of the width.)

(iii) Find the area enclosed by the boundary for a width of 20 metres and for some other different widths.

(iv) Draw a *sketch* graph to show how the area enclosed changes as the width of the boundary increases through all possible values. (Consider both small and large values of the width.)

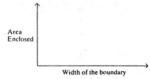

The campers are interested in finding out what the length and the width of the boundary should be to obtain the greatest possible area.

(v) Describe, in words, a method by which you could find this length and width.

(vi) Use the method you have described in part (v) to find this length and width.

More sample assessment activity pages from *The Language of Functions and Graphs, An Examination Module for Secondary Schools,* Shell Centre for Mathematical Education

OPEN-ENDED QUESTIONS

Open-ended Questions

An open-ended question is one in which the student is given a situation and is asked to communicate (in most cases, to write) a response. It may range from simply asking a student to show the work connected with an addition problem to complex situations that require formulating hypotheses, explaining mathematical situations, writing directions, creating new related problems, making generalizations, and so on. Questions may be more or less "open" depending on how many restrictions or directions are included.

Examples:

Open-ended questions help match assessment to good classroom questioning strategies. Here are some examples:

For Grades 1- 4

	AGE									
Don										
Mary										
Steve										
Patty										
Greg										
	1	2	3	4	5	6	7	8	9	10

Look at this graph. Explain what the graph might mean.

For Grades 4 - 9

Luke wants to paint one wall of his room. The wall is 8 meters wide and 3 meters high. It takes one can of paint to cover 12 square meters, and the paint is sold at two cans for $9. What else does Luke need to consider? Make a plan for this painting job.

For Grades 6 - 12

A friend says he is thinking of a number. When 100 is divided by the number, the answer is between 1 and 2.

Give at least three statements that must be true of the answer. Explain your reasoning.

Advantages of Open-ended Questions

There is a wealth of information to be gained from this kind of assessment. The variety of acceptable thinking reflected in student responses goes far beyond what may be imagined. Misconceptions can be detected. We learn whether students can:

- recognize the essential points of the problem involved
- organize and interpret information
- report results in words, diagrams, charts, or graphs
- use appropriate mathematical language and representation
- write for a given audience
- make generalizations
- understand basic concepts
- clarify and express their own thinking

OPEN-ENDED QUESTIONS (continued)

Evaluation of Student Writing by a Classroom Teacher

For a classroom teacher to read all of the papers generated by frequent writing in mathematics might seem burdensome. Teachers who have had their students write, however, say that the results are worth it because they learn more about student understanding and about the gaps in their knowledge. Students have a chance to show more of what they do know, with a wider range of problem approaches.

A random sample of papers can reveal what the class as a whole understands. Give careful review to only part of the individual papers on any particular day, making sure that every student has at least one paper reviewed in a given time span, such as a week or two weeks. Be selective about what is commented on, choosing one or two aspects for detailed feedback.

Holistic Scoring

One of the simplest yet most effective ways to grade student mathematical writing is to use a holistic scoring method. For example, papers may first be sorted into stacks that might be labeled "missed the point," "acceptable," and "has some special quality." Those stacks can then be divided again into two levels each. This method encourages looking for students' thinking rather than small bits of knowledge, and it minimizes the need for structuring questions to elicit predetermined answers.

Rubrics

A "rubric," or a description of the requirements for varying degrees of success in responding to an open-ended question, may be pre-defined or may be created as a result of reviewing a number of papers. If pre-defined, it should allow for the unusual responses that are often seen in open investigative work by students.

Holistic evaluation of papers has been used in scoring open-ended test items by the California Assessment Program. See pages 18 and 19 for more information, or see the CAP report, *A Question of Thinking*. More information on holistic scoring is included in (see bibliography for all): Archbald and Newmann's *Assessing Authentic Academic Achievement in the Secondary School*, White's *Teaching and Assessing Writing*, and the CAP report *Writing Achievement of California Eighth Graders: A First Look*.

> *Judge the piece of work as a whole and give it a grade…. Just get a 'feeling' for the piece of work. Ask yourself such questions as:*
> > *'Does it contain some outstanding feature?'*
> > *'Does it show a spark of originality?'*
> > *'Has the problem been extended (in an unusual way)?'*
> > *'Does the presentation have overall coherence?'*
> *You may be surprised but encouraged to learn that groups of teachers working independently in this way on the same scripts had a considerable measure of agreement on the grades they awarded. (Pirie, Mathematical Investigations in Your Classroom, p. 64)*

This is a page from *Assessment Alternatives in Mathematics*, a booklet from the California Mathematics Council and EQUALS.

SAMPLE OPEN-ENDED QUESTIONS (and one response) FROM CAP GRADE 12 TEST, 1987-1989

James knows that half of the students from his school are accepted at the public university nearby. Also, half are accepted at the local private college. James thinks that this adds up to 100 percent, so he will surely be accepted at one or the other institution. Explain why James may be wrong. If possible, use a diagram in your explanation.

One student's
response:

Imagine you are talking to a student in your class on the telephone and want the student to draw some figures. The other student cannot see the figures. Write a set of directions so that the other student can draw the figures exactly as shown below.

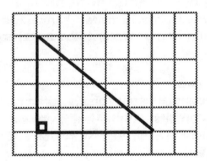 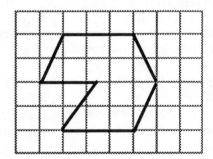

John has four place settings of dishes, with each place setting being a plate, a cup, and a saucer. He has a place setting in each of four colors: green, yellow, blue, and red. John wants to know the probability of a cup, saucer, and plate being the same color if he chooses the dishes randomly while setting the table.

Explain to John how to determine the probability of a cup, saucer, and plate being the same color. Use a diagram or chart in your explanation.

This is a page from *Assessment Alternatives in Mathematics*, a booklet from the California Mathematics Council and EQUALS.

GENERAL SCORING RUBRIC for OPEN-ENDED QUESTIONS
Used for Grade 12 CAP questions

Please Note: For each individual open-ended question, a rubric should be created to reflect the specific important elements of that problem. This general rubric is included only to give examples of the kinds of factors to be considered.

Recommendations: Sort papers first into three stacks: Good responses (5 or 6 points), Adequate responses (3 or 4 points), and Inadequate responses (1 or 0 points). Each of those three stacks then can be re-sorted into two stacks and marked with point values.

Demonstrated Competence

Exemplary Response ... Rating = 6
Gives a complete response with a clear, coherent, unambiguous, and elegant explanation; includes a clear and simplified diagram; communicates effectively to the identified audience; shows understanding of the open-ended problem's mathematical ideas and processes; identifies all the important elements of the problem; may include examples and counterexamples; presents strong supporting arguments.

Competent Response ... Rating = 5
Gives a fairly complete response with reasonably clear explanations; may include an appropriate diagram; communicates effectively to the identified audience; shows understanding of the problem's mathematical ideas and processes; identifies the most important elements of the problems; presents solid supporting arguments.

Satisfactory Response

Minor Flaws But Satisfactory ... Rating = 4
Completes the problem satisfactorily, but the explanation may be muddled; argumentation may be incomplete; diagram may be inappropriate or unclear; understands the underlying mathematical ideas ; uses mathematical ideas effectively.

Serious Flaws But Nearly Satisfactory ... Rating = 3
Begins the problem appropriately but may fail to complete or may omit significant parts of the problem; may fail to show full understanding of mathematical ideas and processes; may make major computational errors; may misuse or fail to use mathematical terms; response may reflect an inappropriate strategy for solving the problem.

Inadequate Response

Begins, But Fails to Complete Problem ... Rating = 2
Explanation is not understandable; diagram may be unclear; shows no understanding of the problem situation; may make major computational errors.

Unable to Begin Effectively ... Rating = 1
Words do not reflect the problem; drawings misrepresent the problem situation; copies parts of the problem but without attempting a solution; fails to indicate which information is appropriate to problem.

No Attempt ... Rating = 0

PERFORMANCE ASSESSMENT

Students Doing Mathematics

Performance assessment involves giving a group of students, or an individual student, a mathematical task that may take from half an hour to several days to complete or solve. The object of the assessment is to look at how students are working as well as at the completed tasks or products.

An observer or interviewer may stay with the group or make periodic visits. Activities may be videotaped, tape recorded, or recorded in writing by an adult or students. The task might be from any mathematical content area and might involve other subjects such as science or social studies.

Assessment Forms

Performance tasks can be supervised by regular classroom teachers or outside observers. Focusing on a group of students or a single student, the assessment may take many forms, such as:

- presenting students with a problem related to what they are already doing in class, and listening to the responses
- observing what students do and say, watching for selected characteristics, making anecdotal records
- interviewing students during or after an investigation
- collecting student writing, either as it is generated by the problem or in response to an additional question

Advantages of Performance Assessment

Looking at student performance gives information about their ability to:

- reason soundly and raise questions
- persist, concentrate, and work independently
- observe, infer, and formulate hypotheses
- think flexibly, changing strategies when one doesn't work
- use manipulative materials, equipment, calculators, and computers
- work together in groups
- communicate and use mathematical language through discussing, writing, and explaining their ideas in their own words
- use estimation
- detect and use patterns
- design and conduct experiments and investigations
- collect, organize, and display information
- get excited about mathematics

… one can reliably judge scientific understanding by observing student teams in a laboratory. Effective means of assessing operational knowledge of mathematics must be similarly broad, reflecting the full environment in which employees and citizens will need to use their mathematical power. (Everybody Counts, p. 69)

This is a page from *Assessment Alternatives in Mathematics*, a booklet from the California Mathematics Council and EQUALS.

PERFORMANCE ASSESSMENT (continued)

Below is an adapted version of an item from the 1986 National Assessment of grades 7 and 11. It was reported in the booklet, *Learning by Doing: A Manual for Teaching and Assessing Higher-Order Thinking in Mathematics and Science.* Students were given a permanently assembled double staircase four blocks high and some loose blocks. There were not enough blocks to build the staircase ten blocks high. The questions are typical of those that might be asked orally by an examination administrator.

HOW MANY BLOCKS ARE IN THE DOUBLE STAIRCASE?

<u>Work with a partner.</u>
1) Look at the double staircase of blocks. How many blocks are in the staircase?

2) How many blocks would you need to build a similar staircase 10 blocks high? How did you figure out your answer?

3) What is the relationship between a similar staircase of any height and the number of blocks needed to build it?

For copies of the booklet and other information, write to
NAEP
CN 6710
Princeton, NJ 08541-6710

This is only one example of performance tasks. Almost any problem on which students work can include assessment, once we decide what to look for. Some examples:

- -

We have reached into this bag of blocks 6 times and have pulled out 3 red blocks, 1 green block, and 2 blue blocks. If you reached into the bag and pulled out another block, what color do you think it would be? Explain why you think it would be that color. How could you get more information?
- Do students have a systematic way of organizing and recording information?
- Do they relate this problem to other similar problems?
- Are they able to express their ideas orally or in writing?
- Are they able to come up with ideas (other than looking in the bag!) for getting more information?

- -

There are 30 students in our class. The office has given us 144 pencils and 24 erasers as our supply for the year. How can we be sure we will still have pencils and erasers at the end of the school year?
- Are students able to make a plan?
- Can they decide when to use a calculator, and then use it effectively?
- Does everybody in the group participate?
- Do students look at all factors of the problem, or do they jump to conclusions?
- Do students use blocks or other materials appropriately?
- Do they make notes or drawings to check their results?
- Do they recognize and use the complexities of the problem?

This is a page from *Assessment Alternatives in Mathematics*, a booklet from the California Mathematics Council and EQUALS.

OBSERVATIONS (adapted from *Assessing Mathematical Understanding*, Project T.I.M.E.)

Focused Observations

Observation, to be effective and illuminating, and to enable the observer to draw some inferences about the students, should frequently be quite sharply focused. Attention to specific details may lead to unexpected insights into a student's understanding.

Students should be observed both individually and as they work in groups. The purpose of an observation may be for mathematics (How far can students count with one-to-one correspondence?) or for affective characteristics (Does this child's behavior help his learning?).

Student Learning Styles

Individuals - Do the individuals:
- consistently work alone or with others?
- try to help others? in what ways?
- succeed in asking for and getting needed help? from whom?
- stick to the task or become easily distracted?
- become actively involved in the problem?

Student Ideas

Explanations - Do the individuals:
- try to explain their organizational and mathematical ideas?
- support their arguments with evidence?
- consider seriously and use the suggestions and ideas of others?
- attempt to convince others that their own thinking is best?

Communication

Verbalization - Do the students:
- talk for self-clarification and to communicate to others?
- comfortably fill the role of both "talker" and "listener?"
- have the confidence to make a report to the whole class?
- capably represent a group consensus as well as their own ideas?
- synthesize and summarize their own or a group's thinking?

Cooperation

Cooperation - Does the group:
- divide the task among the members?
- agree on a plan or structure for tackling the task?
- take time to ensure that they all understand the task?
- use the time in a productive way?
- provide support for each member?
- think about recording?
- allow for development of leadership?

Manipulatives

Manipulatives - Individually or within the group, do the students:
- choose and use appropriate manipulatives?
- fairly share the handling of concrete objects, especially if there is one set for the group as a whole?
- sometimes use the manipulatives only visually? (e.g. count the red faces of a cube without picking it up)
- appear not to need the actual objects but be able to visualize within themselves? (e.g. can "see" the cube in her head)

This is a page from *Assessment Alternatives in Mathematics*, a booklet from the California Mathematics Council and EQUALS.

INTERVIEWS (adapted from *Assessing Mathematical Understanding*, Project T.I.M.E.)

Assessing Understanding

When assessing mathematical understanding, an assessor, whether teacher or outside evaluator, is trying to get a picture of the student's own thinking rather than whether the student can provide the "correct" answer that the adult has in mind. The interviewer wants to know the depth of the student's understanding. Is the student parroting back memorized responses, or has the student personally interacted with the ideas and incorporated them into his or her own conceptual structures?

Questioning Students

Assessment questioning can be brief and informal as in many typical classroom interactions between teacher and students; or it can occur over a more extended period of time where an interviewer really probes to get at what's going on in the student's mind. Interviewing/questioning for mathematical understanding can be done with individual students or groups of students. An adult may observe the students for a while and then, based on what was observed, intervene and ask questions about what the student is doing or how the student perceives the situation.

Plan for Interviews

The logistics of time, people, and curriculum mean planning is necessary for interviews. While students are working on a problem, project, or investigation, an observer/interviewer may observe and question one group of students, taking notes either during the observation/interview or as soon as possible afterwards. Student interviews may also be done by adults other than the teacher outside of the classroom, or at recess or before or after school.

Question Sequence

The interviewer must first find a level of understanding at which the student is comfortable. It is generally better to start asking broad general questions rather than specific narrow ones. Follow-up questions should gradually become more specific as the teacher tries to "zero in on" what makes sense to the student.

Time for Thinking

An important facet of interviewing is the use of wait time — the time allowed for a student to think through a response or to reconsider a response already made. Real thinking takes time.

Multiplication Example

If we wanted to know, for example what a sixth or seventh grade student knows about multiplication, we might ask the student (with manipulative materials and calculators available) to:

- solve a problem in which multiplication can be used
- explain to a younger student what multiplication means
- give an example of a real life situation where 6 x 8 is used
- explain how multiplication relates to addition and/or division

Formal or Interactive Interviews

Assessment by interview may be formal, in which case questions are prepared ahead of time, leading questions might or might not be asked, feedback may not be desirable, and records are kept. Interview assessment may also be an informal regular part of teaching, with more interaction between teacher and student.

ASKING QUESTIONS

Asking the right question is an art to be cultivated by all educators. Low-level quizzes that ask for recall or simple computation are a dime a dozen, but a good high-level open-ended question that gives students a chance to think is a treasure!

These questions might be used as teaching or "leading" questions as well as for assessment purposes. Both questions and responses may be oral, written, or demonstrated by actions taken. The questions and their responses will contribute to a climate of thoughtful reflectiveness.

Some suggestions about assessment questioning:

- Prepare a list of possible questions ahead of time, but, unless the assessment is very formal, be flexible. You may learn more by asking additional or different questions.

- Use plenty of wait time; allow students to give thoughtful answers.

- For formal assessment, leading questions and feedback are not generally used, although some assessment techniques include teaching during the examination.

- Make a written record of your observations. A checklist may or may not be appropriate.

This is a starter list. You will want to build a collection of your own good questions.

Problem Comprehension
Can students understand, define, formulate, or explain the problem or task? Can they cope with poorly defined problems?

- What is this problem about? What can you tell me about it?
- How would you interpret that?
- Would you please explain that in your own words?
- What do you know about this part?
- Do you need to define or set limits for the problem?
- Is there something that can be eliminated or that is missing?
- What assumptions do you have to make?

Approaches and Strategies
Do students have an organized approach to the problem or task? How do they record? Do they use tools (manipulatives, diagrams, graphs, calculators, computers, etc.) appropriately?

- Where could you find the needed information?
- What have you tried? What steps did you take?
- What did not work?
- How did you organize the information? Do you have a record?
- Did you have a system? a strategy? a design?
- Have you tried (tables, trees, lists, diagrams...)?
- Would it help to draw a diagram or make a sketch?
- How would it look if you used these materials?
- How would you research that?

Relationships
Do students see relationships and recognize the central idea? Do they relate the problem to similar problems previously done?

- What is the relationship of this to that?
- What is the same? What is different?
- Is there a pattern?
- Let's see if we can break it down. What would the parts be?
- What if you moved this part?
- Can you write another problem related to this one?

Flexibility
Can students vary the approach if one is not working? Do they persist? Do they try something else?

- Have you tried making a guess?
- Would another recording method work as well or better?
- What else have you tried?
- Give me another related problem. Is there an easier problem?
- Is there another way to (draw, explain, say, ...) that?

This is a page from *Assessment Alternatives in Mathematics*, a booklet from the California Mathematics Council and EQUALS.

ASKING QUESTIONS (continued)

Communication
Can students describe or depict the strategies they are using? Do they articulate their thought processes? Can they display or demonstrate the problem situation?

- Would you please reword that in simpler terms?
- Could you explain what you think you know right now?
- How would you explain this process to a younger child?
- Could you write an explanation for next year's students (or some other audience) of how to do this?
- Which words were most important? Why?

Curiosity and Hypotheses
Is there evidence of conjecturing, thinking ahead, checking back?

- Can you predict what will happen?
- What was your estimate or prediction?
- How do you feel about your answer?
- What do you think comes next?
- What else would you like to know?

Equality and Equity
Do all students participate to the same degree? Is the quality of participation opportunities the same?

- Did you work together? In what way?
- Have you discussed this with your group? with others?
- Where would you go for help?
- How could you help another student without telling the answer?
- Did everybody get a fair chance to talk?

Solutions
Do students reach a result? Do they consider other possibilities?

- Is that the only possible answer?
- How would you check the steps you have taken, or your answer?
- Other than retracing your steps, how can you determine if your answers are appropriate?
- Is there anything you have overlooked?
- Is the solution reasonable, considering the context?
- How did you know you were done?

Examining Results
Can students generalize, prove their answers? Do they connect the ideas to other similar problems or to the real world?

- What made you think that was what you should do?
- Is there a real-life situation where this could be used?
- Where else would this strategy be useful?
- What other problem does this seem to lead to?
- Is there a general rule?
- How were you sure your answer was right?
- How would your method work with other problems?
- What questions does this raise for you?

Mathematical Learning
Did students use or learn some mathematics from the activitiy? Are there indications of a comprehensive curriculum?

- What were the mathematical ideas in this problem?
- What was one thing you learned (or 2 or more)?
- What are the variables in this problem? What stays constant?
- How many kinds of mathematics were used in this investigation?
- What is different about the mathematics in these two situations?
- Where would this problem fit on our mathematics chart?

Self-Assessment
Do students evaluate their own processing, actions, and progress?

- What do you need to do next?
- What are your strengths and weaknesses?
- What have you accomplished?
- Was your own group participation appropriate and helpful?
- What kind of problems are still difficult for you?

This is a page from *Assessment Alternatives in Mathematics*, a booklet from the California Mathematics Council and EQUALS.

STUDENT SELF-ASSESSMENT

The Gift of Independent Thinking

The capability and willingness to assess their own progress and learning is one of the greatest gifts students can develop. Those who are able to review their own performance, explain the reasons for choosing the processes they used, and identify the next step have a life-long head start. Mathematical power comes with knowing how much we know and what to do to learn more.

The Role of the Teacher

This does not mean, however, that teachers abdicate responsibility. Teachers must still help students understand what is needed, provide lessons or activities to meet their needs, identify ways for students to assess what they have done, set guidelines, and ask questions that will highlight the mathematical ideas that are important. The teacher is the stage manager.

Questionnaires

A simple example of self-assessment is a questionnaire following a cooperative activity or project, asking how well the group functioned and how well the student participated. The questions can focus each student's attention on how he or she performed, and can give the teacher the opportunity to talk with the class about successes or difficulties without having to identify individual behaviors. Some typical questions:

Describe the tasks you did for the group _____

What mathematics did you learn?_____

How does this relate to what you have learned before?_____

What could you have done to make your group work better?_____

What worked well in your group?_____

What new questions did this raise?_____

Journal Writing

Still another self-assessment is daily writing in a journal, responding to such sentence starters as:

Today in mathematics I learned_____

When I find an answer I feel_____

My plan for what I will do tomorrow is_____

Of the math we've done lately, I'm most confident about ___

What I still don't understand is _____

Feedback from Other Students

Students can help evaluate by giving constructive comments on one another's work. Looking at others' work helps students develop their sense of standards for their own performance.

For more questions, see page 24 and 25.

STUDENT SELF-ASSESSMENT (continued)

Self Assessment Develops with Practice

...the student needs practice in the form of multiple opportunities and formats for self-assessing. Especially at the start she needs experiences that will stir motivation and give direction to the self-assessment process by enabling her to see where she is, why she is there, and what she needs to go further.

Modes of self-assessment ...include checklists, short answers, paragraphs or essays, charting of what was done in the assessment and in what way, and directed self-learning followed by one-to-one interviews discussing the correlation between faculty assessment and student self-assessment. (Alverno College Faculty, Assessment at Alverno College, p. 13)

Negotiated Curriculum

An example of an integrated student self- assessment at the high school level is found in the Negotiated Curriculum used in Australia, which involves students with teachers in planning and assessing their learning. Students, with the help of teachers, identify their strengths and weaknesses and plan accordingly. The diagram below shows how students decide where they want to go, what they have to do to get there, and what the final results might be.

This diagram is from *Negotiating a Mathematics Course*, by Colleen Vale. See bibliography for more information.

The negotiation of the course — what the students are going to do to meet their aims — adds to the students' understanding of mathematics. The process of reaching decisions about what they will study and how they will do it, parallels some of the steps in problem solving: gathering information, collating and organizing data, interpreting information, forming options, considering implications, weighing advantages and disadvantages through conversation to reach consensus....Involving students in planning the course means that they are clear about what they are attempting to learn from the course...(Vale, Negotiating a Math Course, pp. 6-7)

Alverno College

A similar program for college students is described in the book, *Assessment at Alverno College*, by the Alverno College Faculty, listed in our bibliography.

For all students, the idea of self-assessment is not new. At any age, students can show surprising insight into what they know and what they need to learn.

A SAMPLER OF ASSESSMENT TECHNIQUES

Let us consider, for example, assessing the understanding of whole number multiplication. What do we want to know? What kinds of information would be most helpful? What is most important? How can we judge success? Do we need a number rating or grade? For what purposes? Specifically, what can we have the students do? What are we going to look for?
Are some of the following questions more important than others?

Do students estimate before computing?

Can students use a standard algorithm? Do they have a choice of several algorithms?

Do they know when to multiply? Can they identify a situation that uses multiplication?

Can they explain the process or thinking involved?

Do they understand the relationship between multiplication and division?

A Typical Multiple-Choice Test Item:

$$59 \times 12 = $$

- ○ 608
- ○ 698
- ○ 708
- ○ 798

To Include in a Student Portfolio:

Write a word problem that would probably involve multiplying 59×12 for its solution.

A Multiple-Choice Item to Check Understanding of the Algorithm:

In the following multiplication problem, what number goes in the ☐ ?

- ○ 708
- ○ 128
- ○ 118
- ○ 108

$$
\begin{array}{r}
59 \\
\times\,12 \\
\hline
\boxed{} \\
590 \\
\hline
708
\end{array}
$$

Observation and Interview:

Draw as many diagrams as you can that represent the multiplication fact
$$12 \times 59 = 708$$
Explain to me what each means.

A Performance Task:

(with blocks, beans, balance scales, tiles, graph paper, calculator, etc. available)

You are going to teach a second grader what multiplication is all about. How would you go about this? What materials would you like to use? Show me what you would do.

This is a page from *Assessment Alternatives in Mathematics*, a booklet from the California Mathematics Council and EQUALS.

for Whole Number Multiplication

Here are more possible assessment problems.
Try them with your students.
From which do you learn most?

An Investigation:

Here is a multiplication fact:

59 x 12 = 708

Create a presentation for the class about other mathematical ideas this relates to in some way. You may work alone or with others, and you may consult with other adults or any book in the library.

A Word Problem:

Which operation would be used to solve this word problem? Explain how you know.

Jenny has 4 different skirts and 6 different blouses. How many possible combinations does she have?

An Open-Ended Item:

Name two numbers larger than 10 that you think can be multiplied together to give the answer of 708. Explain how you decided on those numbers.

The hope for your pupils

is that they will be enabled

*to engage in
real mathematical thinking*

*and to see behind the rules
and rote techniques*

*and appreciate a little
of the connected complexity
and insights*

*which go to building up
a deeper understanding
of mathematics.*

*(Pirie, Mathematical
Investigations
in Your Clasrom, p. 4)*

Student Self-Assessment:

How well do you think you understand multiplication? Is there one part or idea you think you may need to work on more?

A Problem for a Group:

Design a test to find out whether the class understands the relationship between multiplication and division.

CALIFORNIA ASSESSMENT PROGRAM (CAP)

State Assessment in California

Since this booklet is written primarily for Californians, we include comments on CAP, as California's state assessment program is known. CAP is the only compulsory mathematics achievement assessment in California. It is developed by California teachers and mathematics educators to match the goals of the California Mathematics Framework.

Matrix Sampling

Currently, CAP uses matrix sampling. This means that any one student takes only a small part of the test, usually during a single class period. No individual scores are obtained or reported.

Assessment of Problem-Solving and a Balanced Curriculum

The current test (1989) includes many items identified as problem-solving and applications (33% of the third-grade test, 56% of the eighth-grade test, and 70% of the Grade 12 test). It assesses a balanced curriculum, including measurement, geometry, numbers and operations, statistics and probability, patterns and functions, logic, and algebra.

Calculators

Use of calculators is allowed with the Grade 12 test; it is anticipated that, as new assessment is created, availability of calculators will be suggested throughout.

Open-ended Questions

Beginning in 1987-1988, Grade 12 asessment included open-ended questions. Evaluation of the responses used the holistic scoring method discussed on page 17. See the bibliography for information about the report, *A Question of Thinking*. In 1990 or 1991, results of open-ended questions will be reported to districts.

Redesign of Assessments

Assessment for all grade levels will be redesigned by the early 1990's, and tenth grade will be added. Early meetings of the 1988-1995 CAP Mathematics Advisory Committee indicate that several assessment methods other than multiple-choice will be developed, such as open-ended questions at all levels, student portfolios and journals, student interviews and videos of students working in groups and using manipulative materials, investigations, performance tasks, and projects that integrate other curriculum areas.

For more information, contact California Assessment Program, Mathematics P.O. Box 944272 Sacramento, CA 94244-2720

California's commitment to authentic assessment rests on a clear vision of a powerful curriculum built on a proper understanding of the nature of learning and knowledge. Thinking—a knowledge-based, discipline-oriented function—is the centerpiece of our reform curriculum. All students are encouraged to think, engage in real-world problem solving, and share in the rich, challenging curriculum that respects the integrity of the disciplines yet emphasizes the connections among them. (California Assessment Program Staff, Authentic Assessment in California, Educational Leadership, April, 1989, p. 6)

This is a page from *Assessment Alternatives in Mathematics*, a booklet from the California Mathematics Council and EQUALS.

SOME ISSUES IN ASSESSMENT

The Issues

It is not possible in this booklet to include all of the important issues about assessment. Many more questions need consideration by the educational community. Here are some statements or positions to stimulate discussion. They do not necessarily represent the uniform thinking of the Campaign Assessment Committee.

- Standardized tests do not evaluate a broad problem-solving curriculum.
- Closed-response tests do not themselves give us an adequate picture of student capability.
- Limited time for completion should not be a factor in assessment.
- Assessment of all students should allow for their unique modes of learning and should feature their accomplishments rather than their failures, with special attention to students with special needs, such as Special Education students, those who do not speak or read English, or those who speak nonstandard English.
- Standardized tests are unacceptable as the sole means of identifying Chapter 1 participants and evaluating their progress.
- Grading can be detrimental to student willingness to learn and should be replaced with other ways to report progress.
- Assessment should provide opportunities for learning.
- Mathematics curriculum and assessment should often be integrated with other subject areas.
- The place of technology in assessment will change and grow.
- New methods of record-keeping will need to be created to support both teacher and student efforts.
- The cost of assessment should become part of the cost of curriculum implementation.

Standardized Tests

Many educators believe that the dominance of standardized tests, while they do provide inexpensive assessment of large groups of students, may be a factor in lowering achievement in mathematics. California Mathematics Council's analysis of publishers' tests, for example, indicated that they did not provide information about student understanding of graphs, probability, functions, geometric concepts, or logic, focusing instead on rote computation. Computation should be neither taught nor tested in isolation. Emphasis on improving test scores inevitably means that what is tested (computation out of context) will be taught, and that what is not tested (thinking and problem solving) will not be taught.

Serious questions may also be asked about the nature and uses of the norming of standardized tests. When and with what populations were the "successful scores" determined? Is it helpful or necessary to identify half of any group of students as below average and therefore less than successful?

This is a page from *Assessment Alternatives in Mathematics*, a booklet from the California Mathematics Council and EQUALS.

SOME ISSUES IN ASSESSMENT (continued)

Closed-Response Questions

"Pre-answered" test items, including multiple-choice, single word answers, or true-false, do not measure the depth of student thinking. Such tests, even those that profess to assess problem-solving, tend to break learning into small parts and to include only simple, quickly solved problems. Students are forced to match the test-maker's formulation of an answer with the test-maker's formulation of the problem.

Generally, when a battery of tests is used, most arithmetic computation skills are tested thoroughly, but items rarely require a student to reason or choose a problem solving strategy. Although arithmetic skills are sometimes clothed in a "word problem," the choice of operation is usually quite apparent, and students are not required to display mathematical power. (Standardized Tests and the California Mathematics Curriculum: Where do we stand? p. 4)

Timed Tests

The use of time as a factor in testing, whether for standardized tests or for practicing basic addition facts, is an equity issue. Many students suffer from enough anxiety in testing situations to keep them from showing their true understanding. Others have physical handicaps, vision problems, or difficulty in writing. Even the slowest student should have as much time to finish work as he or she needs, especially when being assessed.

Pre-answered timed tests cannot allow for mathematical investigations and for working on problems for which the solution is not immediately apparent. Timing of performance tends to discourage persistence and to promote thoughtlessness and jumping to conclusions. Quickest does not equal most talented. The use of time in standardized tests is mainly for the purpose of creating a "normal distribution" by preventing some children from answering all of the questions. Speed is less important to real-life success than reasoning, accurate analysis of situations, and the willingness to tackle tough problems.

Children who have difficulty with skills or who work more slowly run the risk of reinforcing wrong practices under pressure. In addition, they can become fearful and negative toward their mathematics learning....
Speed with arithmetic skills has little, if anything, to do with mathematical power. The more important measure of children's mathematical prowess is their ability to use numbers to solve problems, confidently analyze situations that call for the use of numerical calculations, and be able to arrive at reasonable numerical decisions they can explain and justify. (Burns, The Math Solution Newsletter, Spring, 1989, p. 5)

This is a page from *Assessment Alternatives in Mathematics*, a booklet from the California Mathematics Council and EQUALS.

SOME ISSUES IN ASSESSMENT (continued)

Students with Special Needs

Students who do not have facility in English or who speak non-standard English have the right to the same high quality mathematics program provided for all children. Assessments should be comprehensible to them and not tests of English proficiency. Translators should be provided if necessary. We hope that different kinds of assessment will allow for "friendlier" ways to see how students are progressing, allowing for their own rates and styles of learning.

The same is true of Special Education students. Many of them have undetected gifts that don't show in multiple-choice situations. They should be able to display, through portfolios, performance tasks, and other active assessments, their authentic achievement.

Chapter 1 Assessment

Recent changes in Federal laws, intended to lessen segregation of students, have ended many years of using normed tests as the sole measure of eligibility and achievement of Chaper 1 programs. It is now required that other methods of assessment be used, particularly those that measure advanced skills.

Students from historically underachieving groups should no longer be denied access to more powerful curricula by being diverted into programs designed solely to improve standardized test scores.

Grades

The long tradition of grading students has a detrimental effect on many students. Low grades seldom inspire any student to become more interested in learning. High-achieving students often study for high grades rather than for interest in learning.

Trying to grade every class on a "Bell curve" distribution is a distortion. Our educational goal of learning and success for every child is undermined by a system designed to label some students unsuccessful and the vast majority mediocre.

Parents need better information about how their children are doing than a letter grade and a test score number. Are their children having opportunities to develop thinking? It should be possible, for instance, to say that a student regularly, sometimes, or seldom shows originality of thought and what might be done to encourage creative thinking.

Open Assessment

Assessment should contribute to learning and should involve students themselves in the process. One suggestion, for instance, is to use open notes and open books along with open (available) calculators. There is no reason to assess recall unless memorization is a high priority of the curriculum. In mathematics, it is almost always a low priority, and ready use of necessary information and tools more closely reflects working situations.

This is a page from *Assessment Alternatives in Mathematics*, a booklet from the California Mathematics Council and EQUALS.

SOME ISSUES IN ASSESSMENT (continued)

Drafts of mathematics work should be discussed with peers and teachers and revised before submitting. Not only does this add to the thoughtfulness of work produced, but it will help students learn more about giving and accepting constructive comments.

Record-Keeping

Record-keeping will change with new assessments. Portfolios, student writing, performance tasks, and observations of student work can't be typed into a computer and averaged. Students themselves should be able to maintain records. More space will be needed by every teacher to collect student work. Checklists, if they are used, should be annotated. Time and space for providing documentation and useful information should be built into plans from the beginning.

Integrated Curriculum

Integration of subject matter will be an inevitable part of new curriculum and new assessment. Throughout this booklet you will find reference to use of language and writing in mathematics. Although investigation of purely mathematical problems will continue, investigation of practical questions will take an increasing part in mathematics education at all levels. Practical questions will involve science, social studies, physical education, study of careers, library research, visual arts, and so on.

Technology and Assessment

The place of technology in assessment will change drastically. Instead of on-screen multiple-choice or maintaintenance of grading records, students will have access to interactive programs that give feedback and new information according to student responses. Programs such as *Geometric Supposer* (see *Technology in the Curriculum*) allow students to explore ideas. Periodically the teacher may want to have students record their findings for review.

Time and Cost of Assessment

Some might say that the cost of and time spent on new assessment techniques will be too much. It can be hoped that assessment and curriculum will be so interwoven that assessment time and the money spent will be as valuable as that spent for any other aspect of learning.

Testing of more authentic competencies in large scale assessments may appear to be too costly, compared to the costs of conventional standardized multiple choice tests. But there is reason to question this assumption if one considers the total costs of each approach that involve test development, test administration, scoring, and reporting. Standardized tests entail a tremendous investment in development of individual test items, with relatively lower costs in scoring. In contrast, more authentic approaches involve substantial scoring costs and lower costs for development. (Archbald, et al, Beyond Standardized Testing, p. 32)

WHERE DO WE GO FROM HERE?
STRATEGIES FOR CHANGE

This booklet is meant to inform, to interest, and to spark debate. It's a beginning, not an end, and the rest is up to you. Here is a list of some things you may want to discuss with your colleagues and try:

In Your Classroom

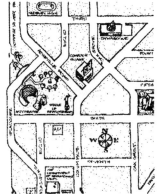

- Look for asessment that matches your goals for your students.
- Get a group of teachers to keep portfolios with you, sharing ideas.
- Give your students some open-ended questions. See pages 17 and 19 for holistic rating ideas that help make evaluation easier.
- Involve your students in mathematical or practical investigations, and talk with the students about how they should be assessed.
- Let your students have a lot of practice communicating orally and in writing about mathematics.
- Take photos of student projects, and put the photos on display with a note about the fact that this is assessment.
- Observe and interview students and make notes about their understanding.
- Videotape your students working on a problem. Have other teachers assess the videotape with you.
- Continue to collect good questions, problems, and investigation ideas.
- Give your students many chances to assess their own progress.

In Your School or District

- Send copies of this booklet and other information to key people in your district.
- Invite administrators, school board members, and others to come and see your assessment program.
- Suspend your present proficiency testing program for a period of time (say 3 years) and try student portfolios, performance tasks, or another method of assessing.
- Seek instructional materials that match the *California Mathematics Framework* and that incorporate mathematics investigations and tasks.
- Investigate alternate possibilities for assessing Chapter 1 students. Check with the California State Department of Education for guidance if needed.
- Inform publishers of commercial tests that you want broader assessment measures than multiple choice.

Anywhere in Education

- Send information on assessment to the media, especially whenever you see test scores reported. For example, organize a student math conference at your school with built in assessment, and invite the local newspaper.
- Let local, county, and state officials know your beliefs about tests and the need for alternatives.
- Continue to look for and develop assessment techniques that will tell us about student understanding and success.

Please keep in mind that all of these suggestions (the whole book, in fact) are presented with the intent of supporting good education for all students. We hope that those who implement any of the ideas will resist making new assessments punitive for either teachers or students. The success of this new adventure will depend upon our trust and belief in each other. We want to make it possible for all students (and their teachers) to show their best work, and to be so proud of their assessment results that they are eager to continue learning.

This is a page from *Assessment Alternatives in Mathematics*, a booklet from the California Mathematics Council and EQUALS.

BIBLIOGRAPHY

CALIFORNIA STATE DEPARTMENT OF EDUCATION:

Write:
> California State Dept. of Education
> Bureau of Publications, Sales Unit
> P.O. Box 271
> Sacramento, CA 95802-0271

California Assessment Program
Annual Report, 1985-86, 1986
> (also available for prior years)

Report of results of annual CAP tests for grades 3, 6, 8, and 12. Gives statewide test scores and examples of many items.

California Assessment Program
Survey of Basic Skills: Grades 3, 6, 8, and 12,
> (4 documents)

Detailed lists of the skills tested by CAP (prior to 1991) at various grade levels.

California Assessment Program, Writing Achievement of California Eighth Graders: A First Look, 1988

Report of results of writing assessment. Reference for holistic scoring comments.

California Assessment Program, *A Question of Thinking, A First Look at Students' Performance on Open-ended Questions in Mathematics,* 1989

Report of results of first open-ended CAP test items, with 20 example problems and scoring rubrics for some.

Mathematics Framework for California Public Schools, Kindergarten Through Grade Twelve, 1985

Statement of philosophy and vision for California mathematics education. Will be replaced by 1991.

Mathematics Model Curriculum Guide, Kindergarten through Grade Eight, 1987

Description of essential learnings, with guiding principles, sample activities, and descriptions of classroom experiences.

Technology in the Curriculum, Mathematics Resource Guide, 1986, and *Mathematics Resource Guide 1987 Update,* 1987, and *Resource Guide 1988 Update,* 1988   (3 documents)

Listing and descriptions of exemplary computer software for mathemtics education.

OTHER PUBLICATIONS

Alverno College Faculty
Assessment at Alverno College, Alverno Productions, 1985

Report of highly developed self-assessment program for college students. Write:
> Alverno College
> 3401 S. 39th St.,
> Milwaukee, WI 53215-4020

Archbald, Doug A. and Fred M. Newmann
Beyond Standardized Testing, Assessing Authentic Academic Achievement in the Secondary School
National Association of Secondary School Principals, 1988

Assessment alternatives, with rationale and discussion. Includes review of uses and limitations of standardized tests. Write
> National Association of Secondary School
>   Principals
> 1904 Reston Dr.
> Reston VA 22091

Assessing Mathematical Understanding (preliminary version by Joan Akers) Project T.I.M.E., 1989

A resource manual designed to help school mathematics leaders plan professional development activities related to assessment. Write:
> Project T.I.M.E., c/o Julian Weissglass
> Mathematics Department
> University of California
> Santa Barbara, CA 93106

Burns, Marilyn, "Timed Tests," *The Math Solution Newsletter* Number 7, Spring 1989

Article on detrimental effects of timed testing of all kinds. Write:
> Marilyn Burns Education Associates
> 150 Gate 5 Road, Suite 101
> Sausalito, CA 94965

Charles, Randall, Frank Lester and Phares O'Daffer, *How to Evaluate Progress in Problem Solving,* National Council of Teachers of Mathematics, 1987

NCTM booklet, with practical tips on how to organize and manage an analytic evaluation program. Write:
> NCTM
> 1906 Association Drive
> Reston, VA 22091

Clarke, David, *Assessment Alternatives in mathematics, The Mathematics Curriculum and Teaching Program*, 1988
Australian professional development packasge for assessment of student learning. Write:
    NCTM
    1906 Association Drive
    Reston, VA 22091

Curriculum and Evaluation Standards for School Mathematics, National Council of Teachers of Mathematics, 1989
Curriculum standards for grades K-4, 5-8, and 9-12, with section on evaluation. Write:
    NCTM
    1906 Association Drive
    Reston, VA 22091

Dossey, John, and Ina Mullis, Mary Lindquist, Donald Chambers,
The Mathematics Report Card, Are We Measuring Up? Educational Testing Service, 1988
Trends and Achievement Based on 1986 National Assessment. Write:
    National Assessment of Educational
        Progress
    Educational Testing Service
    Rosedale Road, Princeton, NJ 08541-0001

Educational Leadership, Volume 46, Number 7, April 1989
Topical journal issue on Assessment.
Write:
    Association for Supervision and
        Curriculum Development
    125 N. West Street
    Alexandria, VA 22314

Evaluation in Mathematics Instruction, Information Bulletin No. 2, 1986, ERIC Clearinghouse for Science, Mathematics, and Environmental Education
Review of research in assessment, prepared by Marilyn Suydam. Write:
    ERIC
    1200 Chambers Road
    Columbus, Ohio 43212

Everybody Counts, A Report to the Nation on the Future of Mathematics Education, National Academy Press, 1989
Emphasizes need for all students to have high-quality education. Write:
    National Academy Press
    2101 Constitution Avenue, NW
    Washington, D.C. 20418

Into Practice: Goal-based assessment and negotiated curriculum, Books One and Two, Victoria Ministry of Education (Australia), 1986 and 1987
Handbook for implementing a negotiated curriculum, with many ideas adaptable to a variety of situations. Write:
    Participation and Equity Program,
    Special Programs Branch, Ministry of
        Education
    416 King St
    West Melbourne, 3003, Australia

Learning by Doing, A Manual for Teaching and Assessing Higher-Order Thinking in Science and Mathematics, National Assessment of Educational Progress, Educational Testing Service, 1987
Hands-on tasks from the national assessment program. For information about this and complete report, write:
    NAEP
    CN 6710
    Princeton, NJ 08541-6710

McKnight, Curtis C., et al, *The Underachieving Curriculum: Assessing U.S. School Mathematics from an International Perspective*, International Association for the Evaluation of Education Achievement, 1987
Review of results of Second International Mathematics Study and their implications. Write:
    Stipes Publishing Company
    10 - 12 Chester Street
    Champaign, Illinois 61820

Stenmark, Jean Kerr, Editor, *Mathematics Assessment: Myths, Models, Good Questions, and Practical Suggestions*, National Council of Teachers of Mathematics, 1991
A brief guide to implementing new assessment programs that contributes to student learning Write:
    NCTM
    1906 Association Drive
    Reston, VA 22091

Phi Delta Kappan, Volume 70, Number 9, May 1989 issue
Major part of journal issue on Assessment. See especially articles by Haney, Neill, Bracey, Wiggins, Rogers, and Raizen.
Write:
    Director of Administrative Services
    Phi Delta Kappan
    P.O. Box 789
    Bloomington, IN 47402

Pirie, Susan, *Mathematical Investigations in Your Classroom*, MacMillan Education, 1987
Guide to implementing investigations in classrooms and for assessment. Write:
    Dale Seymour Publications
    P.O. Box 10888
    Palo Alto, CA 94303

Resnick, Lauren B., *Education and Learning to Think*, National Academy Press, 1987
Review of changes that will be needed in education to promote higher order skills. Write:
    National Academy Press,
    2101 Constitution Ave., NW,
    Washington, DC 20418

Romberg, Thomas A., E. Anne Zarinnia, and Stenven R. Williams, *The Influence of Mandated Testing on Mathematics Instruction: Grade 8 Teachers' Perceptions*, unpublished paper, National Center for Research in Mathematical Science Education, University of Wisconsin - Madison, 1989
Report of research findings verifying that testing does have an effect on teaching. Write:
    National Center for Research in Mathematical
        Sciences Education
    Wisconsin Center for Education Research
    University of Wisconsin-Madison
    1025 West Johnson Street
    Madison, Wisconsin 53706

Shell Centre for Mathematical Education, *Problems with Patterns and Numbers*, Joint Matriculation Board, 1984
An examination module. Classroom materials, with sample assessments. Write:
    Shell Centre for Mathematical
        Education
    University of Nottingham, NG7 2RD,
    United Kingdom

Shell Centre for Mathematical Education, *The Language of Functions and Graphs*, Joint Matriculation Board, 1985
An examination module. Classroom materials, with sample assessments.
    (See address above)

Standardized Tests and the California Mathematics Curriculum: Where Do We Stand? California Mathematics Council, 1985
Review of most commonly used standardized tests and how well they align with the mathematics program described by the 1985 Framework (out of print)

Stenmark, Jean Kerr, *Mathematics Assessment: Myths, Models, Good Questions, and Practical Suggestions*, Natioanl Council of Teachers of Mathematics, 1991
Review and classroom suggestions. Write:
    NCTM
    1906 Association Drive
    Reston, VA 22091

"The Student Writer: An Endangered Species?" *Focus 23*, Educational Testing Service, 1989
Review of some developments in assessment through portfolios and writing. Write:
    Educational Testing Service
    Princeton, NJ 08541-0001

Vale, Colleen, *Negotiating a Mathematics Course*, Brooks Waterloo (Australia), 1987
Description of negotiated curriculum in Melbourne, with students as participants in planning and evaluating their own learning. Write:
    The Jacaranda Press,
    33 Park Road, Milton, Queensland 4064,
    Australia.
    Price is $11.95 (Australian dollars)
        plus airmail cost of $7.45
        (will accept VISA)

White, Edward M., *Teaching and Assessing Writing*, Jossey-Bass, San Francisco, 1985
See especially Chapter 2, pp 18-33, on development and uses of holistic scoring. Write:
    Jossey-Bass, Inc.
    350 Sansome Street
    San Francisco, CA 94104

Wiggins, Grant, "A True Test: Toward More Authentic and Equitable Assessment," *Phi Delta Kappan*, Volume 70, Number 9, May 1989 pp. 703-713
Comprehensive article about assessment alternatives.

Index